Another great book by home improvement expert Tom Tynan that will take you through the process, step-by-step, of installing or having installed energy saving items that will—Save You Money!

In this book, Tom guides you every step of the way as far as tools to use, methods to build or install, and all supplies needed. He says, "Do most of these yourself. Anyone can do it!"

Step-by-Step, ENERGY SAVING PROJECTS, is the book that everyone should read whether they own or rent a home or live in an apartment or condo. These projects are very important because they will save you time, money and increase your level of comfort.

STEP-BY-STEP

with

TOM TYNAN

15
ENERGY SAVING PROJECTS

Swan Publishing
Texas ✳ California ✳ New York

Author: Tom Tynan
Editors: George Tynan & Pete Billac
Layout Design: Sharon Davis
Cover Design: Kenneth George
Tom's Cover Photo: Sonny Schwab

Books by Tom Tynan:

VOL 1 HOME IMPROVEMENT, Homeowner's Most Often Asked Questions

VOL 2 BUILDING & REMODELING, A Homeowner's Guide to Getting Started

VOL 3 BUYING & SELLING A HOME, A Homeowner's Guide to Survival

VOL 4 STEP-BY-STEP, 15 Energy Saving Projects

Copyright @ January 1996
Tom Tynan and Swan Publishing
Library of Congress # 95-72894
ISBN# 0-943629-22-5

STEP-BY-STEP is available in quantity discounts through: Swan Publishing, 126 Live Oak, Alvin, TX 77511. (713) 388-2547 or FAX (713) 585-3738.

Printed in the United States of America.

Dedication

To H.A. Bradley, who showed me the importance of combining my knowledge of building with hands-on applications and taught me to make a dollar doing it. I will always be in your debt.

Introduction

This is the fourth book written by Tom Tynan in two years and he feels—as I do—that this will be his best. Although his first book, HOME IMPROVEMENT, is in it's seventh printing in 21 months and has sold more than 200,000 copies.

His second book, BUILDING & REMODELING, is in it's 4th printing in 14 months and his third, BUYING & SELLING A HOME is being sold all over the world. It is now on CD-ROM and has more than 500,000 copies in print.

STEP-BY-STEP, covers a multitude of energy saving projects, many of which you can do on your own no matter how inexperienced or "clutzy" you are. You will save a lot of money if you follow these instructions. Tom guides you through them slowly and in explicit detail.

"Even if you have work done by a professional it's still worthwhile to read this book," Tom states. "This book can make anyone an expert with the simple instructions given for each project. And you can do most of these projects on your own, even if you know absolutely nothing!"

I am in the same category as many of you when it comes to home repairs—I'm not very good at it. I have, however, become better since meeting Tom. He truly does know his business and how to communicate to even us *clutzy* ones on how to do a variety of home improvement projects.

As his publisher and editor, I can talk about things Tom would not normally mention because along with his truly expert knowledge, his charisma and concern for people, he is unassuming and shy.

For instance, one female listener was so excited over the fact that Tom answered her question so quickly and efficiently, that she went on for a minute or so about how great he is. Tom hears this often. As his editor, publisher and friend, I know he's competent.

For those of you who listen to Tom on his radio show, *Home Improvement Hotline,* or watch his television programs, you know how he gets close to his listeners and viewers. He talks about the repairs he has done and is doing on his own house. He talks about his three young children; about his income tax bill; his dog; his neighbors; and the house he just bought on the beach in Galveston. When he came to Houston in April of 1986, he had a mediocre-paying job at Houston Community College as an instructor, a pregnant wife and less than a thousand dollars in the bank.

Yes, "Things have changed for me since then. Now I have three kids, a house note, and I no longer have even a thousand dollars in my pocket. I'm stretching to buy a beach house," Tom says with a smile. "I work almost all the time and my only vice is my swimming pool and the beach house. I love the water and I swim twice a day, summer and winter."

Tom has an unlisted telephone number but it still rings dozens of times a day. The *get away* house in Galveston is his salvation. He tries to do nothing when he's there, except make tapes on upcoming books and remodel the place. He has to stay busy; it's his nature.

When the day is over, he says, "I sit outside on the deck, look at the water and just relax. The phone doesn't ring."

I don't think Tom is Galveston's most popular citizen, but almost. People recognize him wherever he goes. His voice has been known for years but not until he broke into television did anyone know what he looked like. A few years back, we could eat or talk or visit undisturbed. Now, when in public, heads turn. I can see people whispering and pointing. Many come up and introduce themselves, comment that they like his show or ask a home improvement question.

I think he likes it! Or, he doesn't seem to mind. He answers a question while he's biting into a steak or eating a salad. Some people at his book signings or personal appearances, ask him how to build a *house*—from the ground up! Tom, in his own congenial and caring way, answers each question as though he was building the project himself.

Tom was recently invited to the White House to do a television show for the *Our House* series that is being picked up by television stations all over the country. He left the White House after leaving tips for the president and vice president on how to make that place more energy efficient. His HG-TV series, *Home and Garden Television*, is now shown in more than 50 cities around the United States. Billboards promote Tom and his show in Chicago and Detroit. He is definitely on his way up!

Months ago when shopping for his beach retreat, I

rode with him. This was an experience I want to share because it almost put me in awe of Tom Tynan's knowledge of home repair.

One place he looked at was in slight disrepair. The owner, honest and sincere I believe, was telling Tom the things he would have to do to bring the house up to par.

"You'll need to replace seven pilings but that won't cost much because you can use a backhoe to dig the holes in minutes and use 10-foot pilings. We can have the concrete chipped here, repaired, and . . . ," Tom stopped him.

"No sir," Tom said in a calm voice. "Look at these pilings," he said as he poked his pocket knife under a few of the questionable ones at the base of the concrete. He then rapped with the butt end of the knife against the pilings a few feet up. "You see. These pilings need to be replaced too. Hear the hollow sound up here? I count 15 pilings that need to be replaced."

The owner smiled. Tom said, "Another thing, the height of the piling from the ground to the floor it supports is 7' which means a 10' piling won't pass code. It needs to be as deep in the ground as it is high so we'll need 14' lengths, not 10'." The owner stuttered, "Well, ah, uh, er . . ."

"And," Tom continued, "There isn't enough room for a backhoe to fit under the second floor decking so that will cost more—more time and more expense. Also," Tom went on, "we can't chip the concrete and patch it. It would look awful! We'll have to make a vertical cut here and break up all the concrete and replace it. Also, this section here can't stay or the water will run to this other side. To do it correctly, we'll have to set a grade of this concrete so it slopes *away* from the house . . ." Tom spoke for several more minutes pointing out in expert detail what really had to be done to the place. My question is, "How does he know so much?"

Apparently, the man had never listened to 740-KTRH on Saturday or Sunday from 11am until 2pm or ever watched Channel 11 Saturday from 9:30am to about 10 or on Sunday evening during the five o'clock news, or he'd have known he was dealing with *the expert.*

I stood there, respectful of the way Tom explained the condition of the property, not smug or sarcastic but with the ease in which he tells each of you how to remedy a problem with your home.

This is what is so appealing about Tom Tynan. He listens intently to your question then gives you an answer. He explains and never belittles. He teaches without making you feel dumb. He saves you money with his wisdom. He cares about each question and will stand by each answer. If he doesn't understand or doesn't know, he tells you that too. I agree with that lady; the sonofagun *is* great!

I got a call a few weeks ago from a guy who suggested Tom retract a section in the *Home Improvement* book, about using Teflon tape on gas lines. The man explained Teflon deteriorates and is unsafe. One should use "plumbers putty." I asked the man if he wanted Tom's number because he sounded like he knew what he was talking about. He refused. He didn't even want to tell me his name. I would, however, like to thank him for the call because he meant well.

I immediately called Tom. "Nope!" Tom said. "Perhaps the *old* Teflon was not up to par but there is a new Teflon tape that is approved for gas lines and the manufacturers stand behind it. They use Teflon products on the space capsules and I've been using it for years with zero problems. Have the man call me and we'll discuss it."

For those of you who have never met Tom, it's simply because you haven't tried. He is "somewhere" almost every day either giving safety lectures, appearing at home shows, at one of the many bookstores autographing his books, taping a new television show, doing his weekly radio show, cutting radio commercials for the many advertisers who are waiting in line for him to do their commercial. Other than his broadcasts, personal appearances, etc., you might also see him in any home improvement store looking and learning about any new tools or materials that are introduced.

He is also invited to manufacturing plants by those who would like to have him advertise their products. Before he accepts, he checks them out carefully. Tom has passed up many a lucrative offer because he does not feel some products are the best for his listeners, viewers or readers. He won't steer you wrong. Yes, he's my friend and yours too.

Pete Billac
Editor and Publisher

Table of Contents

Chapter 1

RIDGE VENTS

For those of you who listen to me on radio or watch me on television, you will hear me mention ridge vents at least every other show. I talk about them because they are so beneficial. They have become so popular that whether I plan to talk about them or not, somebody calls in and asks, "Tom, what *is* a ridge vent?" I know from the question this is not a regular listener.

A ridge vent is nothing elaborate. It's a vent that goes across the ridge of the roof from end to end, a cover that fits over a slot that is cut through the roof deck allowing your attic to "breathe" at its highest point.

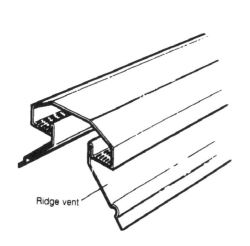

Ridge vent

Basically there are two types of ridge vents: ones that are made of metal and ones that are made of PVC. The metal ones were the original vents, but did not perform very well. The concept was good but there were problems in

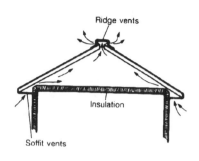

the design. The two biggest problems were that the nails rusted which attached them to the roof because of their exposure to the weather. Also, during a heavy rain they would tend to leak. (They were also very ugly!)

PVC ridge vents are what I recommend. All the nails are protected from the weather by the shingle cap. The PVC vents are also designed to endure winds of up to 140 miles per hour without leaking.

They also look very good and are so low profile that it's not easy to see them unless you really look closely.

During hurricane Andrew, homes that had these PVC ridge vents weathered far better than homes with different types of vents such as turbines or static vents.

To install these vents yourself, I'd recommend only an experienced do-it-yourselfer try it. But, it's easy to learn if you follow what I tell you, step-by-step. The biggest danger is falling off the roof. If you're uncomfortable with height or balance don't do it. If you aren't, here's how it's done.

First, measure your roof along the topmost edge. Then go to your roofing supply company with this measurement. Most of these ridge vents come in four and 5 foot lengths. You will need to get enough ridge vent to cover all your ridges. If your home has more than one ridge, get all the measurements and total them up. This measurement tells you how many you will need.

Bring along a piece of your roof shingle. You will need extra shingles to cover the ridge vent. Also, you will want the color to match your existing roof. The sales person at the roofing supply house will be able to tell you how many bundles of shingles you will need to cover these ridge vents.

INSTALLATION

A *chalk line*, a *power circular saw* with several carbide tip blades, a *hammer, 2 ½" roofing nails*, and a *razor blade mat knife* are required to do the job.

Now that you're on the roof, measure down from the peak about an inch or inch and a half on either side. Do this at both ends of the ridge. Take a chalk line and snap a line on both sides of the peak. Use this mark to make your cut.

Simple to follow instructions in the ridge vent packages will tell you how big a slot you need to cut.

Now, set your circular saw blade to the **depth of the deck** and begin cutting along that chalk line—about a foot in from each end. Be careful not to set the blade too deep that it cuts through the rafters. My "safety" rule is to never exceed an inch in depth. It won't do too much harm to cut a little into these rafters but I've seen people not set the depth and cut more than halfway through a rafter!

Most roof decks have plywood averaging a half inch in thickness. The shingle on top is maybe another half inch. The thing to do is cut a small section, see how deep it is cutting and adjust the saw blade accordingly.

When making this cut, plan on cutting through nails, shingles and roof deck. This will dull the saw blade very quickly. Please, wear safety goggles. Debris will be flying, the saw will begin to smoke and most first-time ridge vent installers will have a mild heart attack.

When the blade gets dull, it tends to smoke and stop cutting. This is the time to change to a new blade. The price of a few extra blades makes the job a lot more palatable.

Now, after you've cut that slot on each side of your roof clear it out. Just for grins, put your hand over the open slot and you'll immediately feel the hot air escaping. The slot will not be perfectly straight; it's a tough cut. But don't panic, the ridge vent will cover this slight imperfection and it will look terrific. This isn't a TV cabinet or bookcase you're building, so expect a little wiggle and waggle.

You're ready to place the ridge vents on one at a time and begin to nail. The first thing you want to do now is pop yet another chalk line where you want the edge of your ridge vent to be. This will make the vent nice and straight. Although you cut the hole a foot in from each end, still run the ridge vent to the end of the roof because it looks better. Another caution is to not *overdrive* the nails. You will risk smashing the plastic on the ridge vent.

All ridge vents have a *nail line* drawn on the top of the vent; some even have pre-drilled holes. Put your nails as they indicate. This will be shown in the instructions.

The final step is to cover the ridge vent with a shingle cap. Nail them down normally or the way you'd do a normal ridge cap. Make certain you nail through the nail line that is stamped on a ridge vent.

> Don't be frightened by this do-it-yourself project; it really is easy to do. I started this book with this subject because it is so important to your energy savings.

If you are having a new roof installed don't neglect getting ridge vents. If your roof is five years old or less and you don't have ridge vents, I think it's wise to put them on. You see, a ridge vent not only lowers your energy bill but extends the life of your roof. It exhausts the moisture and heat which tend to prematurely destroy shingles.

With *soffit* vents and ridge vents the temperature in your attic should be about the same as the outside temperature, give or take a few degrees. When the sun is beating down at midday, the temperature should not be more than 10° hotter than the outside temperature. Unvented attics can reach a temperature as high as 150° at this time of day.

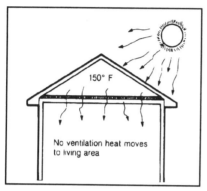

No ventilation heat moves to living area

Remember too, that proper ventilation is just as important in the winter as it is in the summer because you *exhaust* the moisture. And a moisture build up during the winter

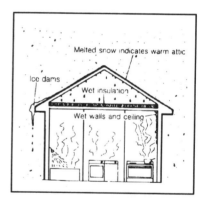

Melted snow indicates warm attic

Ice dams

Wet insulation

Wet walls and ceiling

can reduce the effectiveness of the insulation.

The drawing to the right shows snow and ice on the roof. However extremely cold temperatures can produce the same effect.

Chapter 2

RADIANT BARRIERS

One of the most misunderstood energy saving items in existence is the **Radiant Barrier**. This is what it does. The Radiant Barrier **stops** the radiation from the sun from being transmitted into your home to heat things up.

Have you ever noticed when you come home in the evening after the sun goes down that it's literally hotter **inside** your home than it is on the outside? That's because, all day long, your home acts like a *sponge* that soaks up and stores all this radiant heat gain.

This heat gain makes your air conditioner work harder because it now has to **remove** the heat that is retained in the house. It plays havoc with your energy bill. A radiant barrier stops your home from soaking up all this unwanted heat.

Radiation is often misunderstood. A good example follows: When you stand in front of a fire on a cold winter night the part of your body facing the fire feels warmer than the part of your body that isn't. That's because that part of your body facing the fire **absorbs** the radiation from the fire. Radiation heats an object, not the air.

Another example (maybe a better one) is that you feel cooler under a tree when the sun is shining. The air temperature is still the same but the radiation from the sun is not beating down on you. A radiant barrier is a sort of shading device. If you take a piece of aluminum foil and cover your body with it, it protects you from absorbing radiant heat from the sun or a fire.

The sun beats down on your roof; your roof heats up; it transfers this heat into the attic. This heat then reaches the next thing that is cool which is your insulation. As your insulation heats up it radiates that heat down into your house.

If you don't believe me, when the sun goes down, go touch your insulation. It will still be hot because it has absorbed all that radiation from the sun. A **Radiant Barrier** in a home is simply aluminum foil (as opposed to gold foil which is what they used on the lunar spaceship because it's an even better barrier than aluminum but more expensive).

They make this foil in rolls for builders to use. This aluminum foil is stapled to the bottom side of the rafters in your attic and stops 95% of the radiation from the sun. It works well in conjunction with properly ventilated attics which will work in any climate. In southern climates or wherever there is a lot of sun, you need a radiant barrier. If you live in a cooler climate or you're not worried about the building absorbing heat you would not need a radiant barrier.

What you're more concerned over in a cold climate is in trapping the heat **inside** the building. This is where thicker insulation is more effective than a radiant barrier.

There are things the homeowner needs to be aware of when purchasing radiant barrier material. First, some of it is too expensive. I'd suggest paying between 10-12¢ per square foot. Some salespeople might try to sell you DOUBLE SIDED at a higher price, SINGLE sided is all you need.

Since one sheet stops 95% of the radiation, a second sheet will only stop 95% of that remaining 5% so it isn't worth it. Don't let anyone tell you differently.

Some companies and salespeople will say you can lay the barrier material on the FLOOR of your attic. NOT SO! When a radiant barrier gets dirty, it stops working. Also, by laying it on the floor of your attic over your existing insulation, it will tend to trap moisture below it and reduce the effectiveness of the insulation.

INSTALLATION

It is simple to install a radiant barrier on your own. The first thing to do is measure how much you need. Radiant barrier material is sold by the square foot. Go up on your roof and measure. It's easy. Just get the length and width of each roof plane and total them up.

The tools needed is a razor blade mat knife, a staple gun, a lot of staples and a tape measure. A

POWER staple gun makes the job much easier. They can be bought for about $30. Using one will save your hands. With a hand powered staple gun your arm will go numb after the first hundred squeezes.

Face the shiny side DOWN. Let it face the air space. It would still work if you face it up but chances are it would get dirty. When it gets dirty (remember) it doesn't work. Face the aluminum DOWN to where you can see it as you staple. Remember, it doesn't have to face the sun because it doesn't reflect the heat, it's a barrier between you and the radiation. As long as it faces the air space (the attic) it will work!

Radiant barrier material comes in a giant roll. I'd suggest smaller sheets, like maybe 10' sections cut with your razor mat knife. Work one section at a time. Don't cut any framing members (2x4's) or block attic ventilators. If it rips and tears a bit, that's okay. Radiation moves in a *straight line* and won't move around searching for an opening to penetrate.

Most attics have areas that are difficult to get to such as the part where the rafters meet the outside walls. That's OK. It isn't necessary to get it all. If you get 80% covered, you'll stop a significant amount of radiation from entering your building. Of course, it's easier to have it put up when the house was being built. In fact, you can now buy plywood with a radiant barrier material applied to one side. This type of plywood is being used for roof decking, thereby achieving two things at once: decking and the radiant barrier.

If you use a radiant barrier you will need *less* "R" value on your insulation since it stops the majority of the heat radiated by the sun. For instance, a radiant barrier with "R" 19 fiberglass insulation will outperform the same house with just an "R-30" fiberglass insulation.

ANYWHERE the sun beats down on the roof a radiant barrier is necessary and wise. If you have trees shading your roof, forget the radiant barrier and only go with insulation.

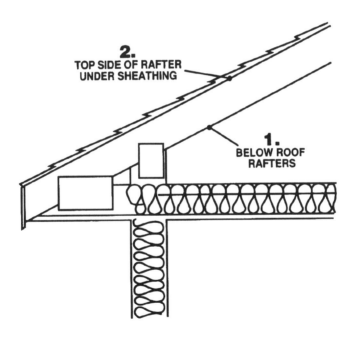

Recommended Locations for a Radiant Barrier

Chapter 3

ATTIC INSULATION

If you want to *really* save on energy bills, insulating your attic is of great importance. Most homes do not have proper levels of insulation. In the winter heat rises from the interior of your home through the attic to escape. In the summer, heat (the sun) beats down into the attic thereby heating the inside of your home. It's backwards in regards to saving energy and money, isn't it? That's why there is need to have more insulation in your attic than in any other part of your house.

This is where we get into the "R" value. I explained it pretty well in my first Book, *Home Improvement— Homeowner's Most Often Asked Questions.* Now I will go into greater detail.

The "R" stands for *Resisting Heat Transfer* and it is gauged by number; the higher the number, the more heat transfer it will resist. When you talk about insulation for example, the thickness determines the "R" value and thus it slows the heat transfer that begins with the sun beating down on your roof, through your attic, and into your home. The "R" value of the insulation slows this heat transfer considerably, from your attic to your living area.

Some people buy insulation and make the mistake of packing it down really tight. If you buy an R-30 insulation, make sure you let it fluff up to its full thickness or it won't perform as an R-30. It will not be as effective. The thickness and the "R" value determine the effectiveness of the insulation.

The first step in insulating your attic is to find out how much you need, if any. Take a measuring stick to the attic, hold it above the insulation and wiggle into it until the stick touches the attic floor. Mark it and check the thickness of the insulation you already have. Call your local power company with this information. They will check their chart to tell you what your "R" value is. The following chart is a good example of loose fill fiberglass insulation. It shows the "R" value and the corresponding thickness.

R Value	Minimum Thickness
R-49	19½"
R-44	17½"
R-38	15¼"
R-30	12"
R-26	10¼"
R-22	8¼"
R-19	7½"
R-11	4½"

For instance, if you find you have 4" of fiberglass insulation and your power company recommends 12", you will have to add an additional 8"; it's that simple.

Since the information in this book applies to any location in the United States (anywhere in the world, really) the very **best** way to determine the "R" value is to make a telephone call to your local utility company. In southern regions it's around R-30. Northern regions could go as high as R-42.

Before you go out to buy any insulation, you will have to know how much to buy. There is no need to crawl through your attic to measure. If you have a two-story home, measure the square footage of each room *below* the attic. If you have a single story home, measure from the outside. Easy enough, huh? Do *not* measure garages or porches; only rooms that are air conditioned or heated.

> For instance if one room measures 10'x12', this totals 120 square feet. Measure every room then add the total. If your room is 10'7", round up the number to 11'. You will need the extra anyway.

TYPES OF INSULATION

▶Fiberglass—It's come a long way in the last 20 years and in my opinion, is *still* one of the best options for homeowners. The reason I like it is because it does not deteriorate over time. Also, Fiberglass Batts (or blankets) is the easiest method for do-it-yourselfers to insulate an attic.

▶Cellulose—Environmentally friendly because it's made from ground up newspapers but they have to add a boric acid solution to give it fire retardant properties. Many people are allergic to this type of chemical. Too, over a period of time it tends to decompose.

▶Rock Wool—It's been around since the late fifties and early sixties. The problem is that it tends to break down easily and scientists are constantly trying to upgrade it. It's also very abrasive and expensive! My choice is FIBERGLASS!

If you follow my recommendation, and choose fiberglass, you have two choices for installation:

☜Have it **blown** in by a contractor who has a large blowing machine. That's the easiest way and you'd be amazed at how inexpensive this is to have done. Remember, call at least **three** contractors. Tell them your square footage and the "R" value and ask them to place a bid on what it will cost to bring that "R" value up to the recommended level.

☜For you do-it-yourselfers, rolls of fiberglass insulation are what I recommend. Here's the step-by-step process.

The first thing you need to be aware of when you buy insulation is you're adding insulation on top of existing insulation, so you must purchase **unfaced batts**. In other words, no paper or foil on either side of the insulation regardless what the salesperson tells you.

TOOLS

1. Razor Blade Mat knife.

2. Straight edge (a yardstick will do).

3. Mask, goggles and gloves. Wear an OLD long-sleeved shirt to cover your arms and neck. (When you've finished, chances are you'll want to throw it away).

4. Tape measure. And, that's it.

Most of these fiberglass insulation rolls come in 4' lengths but the widths are different. If you are insulating between your ceiling joists, make certain you buy the width that will fit between them. If your present insulation is at the top of your ceiling joists, neither the length nor width makes a difference because you'll be rolling across the top of these joists.

This fiberglass insulation is easy to cut. Lay it on a flat surface, like a board. Put the yardstick or straight edge along the line you want to cut, mash it down so it's tight against the board, then cut it. Take a couple of *swashes* through it with the blade and it will cut nicely and clean. Cut it two or three inches longer so you can "squish" it in reasonably tight. Don't throw away any scraps. You might want to use them in areas that need extra.

Try to work on your attic insulation in the cool weather, not when it's a 100° outside; it gets mighty hot up there. If it's summer when you do this, make certain you do it in the early morning or evening when it's cool. It doesn't all have to be done in one single day either. Do some, take a break, do some more, go to sleep, go to work and when it gets cooler start again. It's easy to do and you will save many dollars if you do it yourself.

Here's a tip too. Don't **unroll** the insulation until you get it to your attic. I know it *seems* easier to unroll it in your driveway and then cut it but not so; when you cut the package open it will expand several times its original size and chances are you'll never be able to *get* it in your attic. Also, you don't want to drag all that fiberglass through your house. Keep it rolled, carry it to your attic and then unroll it.

Some attics have rather low points (especially where the roof meets the outside wall). There is no need to try and crawl up under it. Put that insulation down and use your yardstick to sort of push it flush. Be careful not to push it in so tight that it stops the air flow from your *soffit* vents.

This isn't an *expense.* It's an *investment.* You'll be paid back by energy savings on your utility bill in three to five years, depending on the size of your home and how much insulation you had to add.

Chapter 4

SOFFIT VENTS

We all take breathing for granted. We breathe air in and we breath air out. Homes do likewise. Your home is built in the shape of a box. Consider your roof as a big hat mounted on top of the box with a 2-foot airspace between this *roof-hat* and the box which will keep the wind, rain and sun from ever hitting the box (your house).

I will start by telling you what a *soffit* vent is. The *soffit* is the bottom board of the overhang of your house. If you walk around the outside of your house against the wall and look straight up to where your roof hangs over, this overhang is called a *soffit.* Soffit **vents** only go on the **lower** overhangs, not on the gable overhangs. Better yet, here's a drawing of a *soffit* vent.

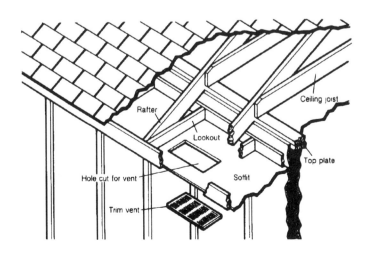

The *soffit* is simply the board where the vents are mounted. Your **upper** roof vents, (ridge vents or whatever they may be) will not work unless they are "fed" by cooler air. The closer you are to the ground the cooler the air. When this cool air heats up in your attic it rises and escapes out the upper roof vents only to be replaced by more cooler air through the *soffit* vents.

All homes **need** *soffit* vents. They are easy to install if your home does not have them. It's important that you have them **all around** your home, not just on one side.

Most *soffit* vents come in two types: one is a 4"x16" grill and the other, an 8"x16" grill. The 4"x16" type must be installed on 3' centers. The 8"x16" must be installed on 4' centers.

Most might reason the larger grills are twice as big as the smaller type so they can be twice as far apart. Not so! If they are too far apart, they will leave a **stagnant air space** between them. **Even** air flow is as important as the **volume** of air.

To install *soffit* vents, let's first determine how many you need and the type you'll use. I like the 4" type better because I think they look better even though you have to install more of them. The cost is almost the same as the larger ones but you don't need as many. I also like the smaller ones better because *fewer* holes have to be cut. Look at both and choose. If your overhang is not wide enough for the larger ones, you have no choice!

Now, let's assume you've chosen to install the larger vents, the 8"x16" grills. After you measure the length of the overhang plan to put these larger grills on 4' centers. This means that you divide the length by four (round the number off) and that number is how many of these grills you need. For example a 21' length divided by four equals 5.25 vents. Round this number off to five vents. Simple enough, huh?

Pattern with 3" Hole Saw (4"x16")

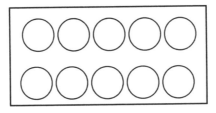

Pattern with 3" Hole Saw (8"x16")

Now, let's put them up. The first step is to estimate where these vents will go and lay them out accordingly. Always start from the center of the overhang and work your way to either side.

If you're installing an **odd** number of these vents (5, 7, 9, etc.) your first vent will go in the very *middle* of the 21' distance. (See diagram page 38.)

If you're installing an **even** number of vents (4, 6, 8, etc.) follow closely. Remember I said they are to be on 4' centers. Well, let's assume your overhang is say 32' and you are putting up eight vents. Still go to the center and mark a line on that 16' center (half of 32 feet). Now comes the tricky part. Measure from the **center** line **exactly** 2' on each side. Then, place the *center* of the vents on these lines. (See diagram page 38.)

To make certain they all line up straight, you will need a *chalk line*, a *tape measure* and *pencil*. Measure back from the front of the *fascia* board 3-4", depending on the measurement of your overhang. Most overhangs are between 16-24" out from the wall. If your overhang measures differently you will have to use your own judgment. Just make certain you measure *a few inches in* from the *fascia* board so you'll have room to work. Do this at each end of the overhang.

If you have a helper have him or her hold the chalk line at one end. If you are working by yourself, put a nail on the measurement on one end and tie the chalk line to it. Then go to the other end and hold that chalk line over that measurement and snap the line thereby giving you a straight line.

The absolute *easiest* way to cut the holes for these *soffit* vents is with a standard 3/8" or 1/4" drill (everybody

has one of those) and a hole saw. If you're not familiar
with the hole saw, it's not really a saw but an attachment
that fits on the end of your drill. I recommend that you get
a 3" hole saw. This size will work with either the 4" or 8"
soffit vents. You can buy them at almost any hardware
Store or home an improvement center.

Hold the grill up and make a pencil outline of the
vent on the soffit. Then take your drill (with the hole saw
attachment) and drill a series of holes *inside the outline* of
the vent, just make certain you're at least a half-inch
inside this mark so you will have wood to screw into.

When you install these soffit vents, have the grill
facing to the *inside* of the house (so you can't see into it
from the yard) and with a screwdriver and screws, secure
it in place.

After you make your lines and layout, it's relatively
easy. When you finish installing a vent, take a wet sponge
and wipe off the chalk line. I'd use the same paint you
painted your house with to make the soffit vents blend
unnoticeable.

Once you do one, the second and third, etc., will be
a snap. Don't be afraid to try. It truly is an easy job, even
for rank beginners.

To those of you who noticed in the picture of the
soffit vent on page 33, there is a rectangular opening cut.
This is for professionals. I mentioned the *hole saw* is

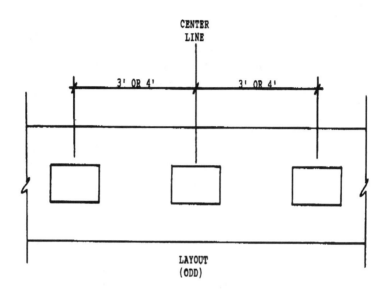

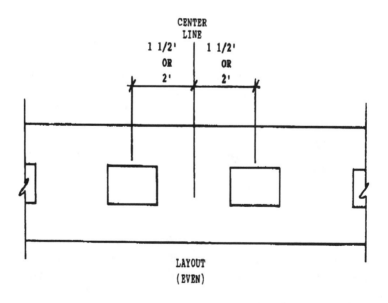

easier for most do-it-yourselfers because it fits in the end of a drill. To cut a rectangular hole do NOT use a circular (*Skil*) saw and hold it upside down; it's too dangerous! The professionals use what is called a Reciprocating Saw (*Sawzall*) that has a long blade used to cut rectangular openings. Stick with the round holes. They work well. This

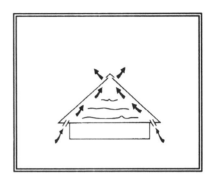

drawing depicts perfect air flow when you have ridge vents and soffit vents.

Chapter 5

CEILING FANS

The ceiling fans of today are not only functional but also attractive pieces of furniture. Many people use them in every room in the house. I'm going to tell you **how to use them properly** and to save money on your electric bill.

In the summer, set fans on a minimum of MEDIUM SPEED and make sure the little switch located on the body of the fan, below the blades and above the light kit, is set for BLOWING DOWN! Most ceiling fans use about as much electricity as a 100-watt light bulb, which is next to nothing.

Most of us have little choice as to where the ceiling fan is located, usually in the center of your room. If this is so, try to position the furniture so you'll be sitting directly underneath them. You will immediately feel 4-6° cooler than whatever the thermostat reads. Therefore, you can turn that thermostat UP 4-6° which will save a substantial amount on your electric bill.

In selecting a ceiling fan make certain it is large enough. (See chart on the next page) I know there are bargains and fans are advertised on sale all the time. Just make certain they are REVERSIBLE. (Fans that can blow the air down or up.) Next, choose one with FIVE blades as opposed to three or four. More blades move more air.

Dimensions of Room	Minimum Fan Size
12 ft. or less	36 inches
12 -16 feet	48 inches
16 -17 feet	52 inches
17.5 -18.5 feet	56 inches
18.5 feet or more	2 fans

A chart for determining what size fan is needed for your room.

You also want your ceiling fan placed far enough from the ceiling to work as well as possible. For instance, most ceiling fans come with a short pole that are designed for the standard 8' ceiling. The newer homes have ceilings 9' or 10' even 12' high. A *Cathedral* ceiling might be 20' above the floor. In that case, you will have to get an extension to bring the blades down to about 7-8' from the floor. If you mount it any higher, it simply won't move the air efficiently enough to cool that particular area.

In the winter, REVERSE the fans and put them on LOW speed. You see, as your heater warms up your house the hot air rises to the ceiling and the cold air hugs the floor. That's why your feet are always cold because your feet are in the coldest air-layer in the house.

In turning these ceiling fans in reverse, they tend to circulate the air, thereby mixing the hot with the cold which prevents the air in the room from becoming stagnant; it stops air from stratifying. At the same time you don't feel the air moving down across your skin like in summer, therefore you don't feel a draft.

Since the thermostat on your heating system is mounted low on the wall, like maybe 4-5' high, it will read cooler than it really is because the lower air is cooler.

INSTALLING A CEILING FAN

All new ceiling fans come with instructions on installation but very few have diagrams on *mounting them* in different situations. Each fan is connected to the ceiling a little differently. But there are two things every ceiling fan has in common. One is how they are wired and the other is ceiling fans are heavy and have to be secured to the framing of the house.

SECURING THE FAN

Most electrical boxes are not connected to the framing; they are meant to hold light fixtures—not heavy ceiling fans. Therefore, the first thing to do is to find out if the light box is connected to the framing. If not, you must find some way to support it. The only way you'll know for sure is to go up in the attic and look.

If it is not attached to the framing, get a 2x4 and cut

it to fit between the ceiling joists. Place the 2x4 between the ceiling joists so it rests on the back of the electrical box. Next, nail through the ceiling joist into the 2x4 using framing nails.

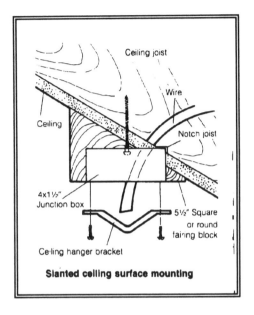

Slanted ceiling surface mounting

Now, get out of the attic and from inside the room screw the electrical box to the 2x4. At this time you're ready to open the kit and follow the instructions provided.

WIRING

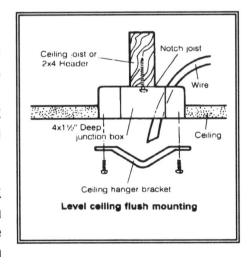

Level ceiling flush mounting

With light fixtures or ceiling fans, wiring is always the same. First, turn the power OFF at the breaker before you work with electricity.

You'll see a black wire, a white wire and a bare (or green) wire coming out of the fan AND out of the electrical box. If the fan has a light kit, there is also a blue wire.

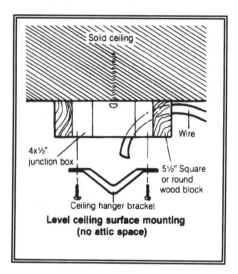

Level ceiling surface mounting
(no attic space)

Anytime you work with wires the white wires go together (one from the fan pairs with the one from the electrical box), the black wires pair together and the green or bare wires go together; this is the ground. The one exception is with a light kit the blue wire connects with the black wires. It's that simple.

If the electrical box is made of plastic, you must change it to the metal kind approved for ceiling fans!

I recommend you enlist a helper for fan installation, preferably one that is reasonably strong. You can do it by yourself but only if *you* are reasonably strong. It's not easy trying to balance the workings of the fan with one hand and connect the wires with the other.

There is a product now that you can get at your local home improvement center that is called an EASY FAN BRACE that makes installing a fan easy and literally keeps you out of the attic.

If you can't or don't want to run a power cable for a

ceiling fan or want to install one in a garage, workshop or hunting lodge, this is an easy way to do it. Secure it well. If you don't, it may vibrate loose and fall on your head. You'll remember to do it the next time.

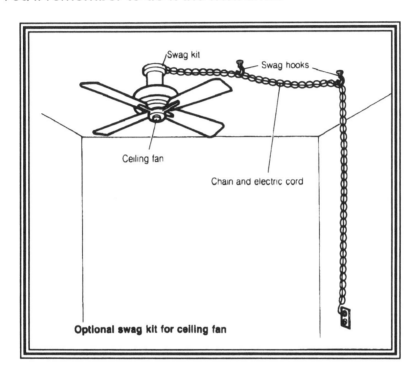

Optional swag kit for ceiling fan

Chapter 6

CAULKING

Most people don't realize infiltration (air coming into or leaving their home) can cost a lot of money on an energy bill. This can happen in many places around your home *especially* around your windows and doors.

Let's start with windows. FOR ENERGY SAVINGS, remember to caulk around the **inside** of your window frame, not the window pane itself and **not** the outside! You can caulk around the outside where the window meets the brick or wood siding so rain doesn't penetrate into the walls.

As far as energy savings it's the **inside** of the house where the window frame meets the sheetrock or the wood. You want a tight seal so your home won't be *sucking in* heat during the summer or allow heat to *escape from* the home during the winter.

This is simple to do. Go to your nearest home improvement center and get a tube of **Acrylic Latex Caulk** *with* Silicone! Get three or four or five tubes; they're only about three bucks a tube. You will also need a quality caulk gun. A good caulk gun is a good investment, dependable and is easier to use. You can also use it on painting projects as well as caulking windows.

I planned to show drawings of these various tools but decided against it. I would prefer you take my book to the home improvement store and look at the tools I've listed. This way you can see it, feel it, test it and ask the salesperson for additional information. Besides, I don't want to make it too easy for you.

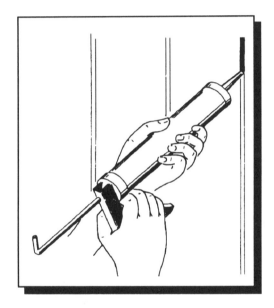

Besides the caulking gun, another tool is a good putty knife with a 1" wide blade. Check the windows and see what needs to be done. Scrape away any loose caulking with your putty knife. Pour a little mineral spirits on a rag and wipe down the area you plan to caulk; this will help the new caulk adhere better. You'll also need a sponge and a bucket with water.

Load the caulk tube into the caulk gun. Cut the plastic tip off for a hole about an ⅛" round which is a little smaller than the size of a pencil. It is no longer necessary to poke a hole through the inner seal of the tube. In the last several years these new acrylic tubes have been repackaged. Make the cut at about a 45° angle (scissors will do).

Squeeze the caulking out slowly as you steadily pull the trigger on the caulking gun. It will look pretty rough but that's okay, apply it to the window frame/sheetrock surface as evenly as possible. Then take a damp sponge and gently press and drag it over the caulk bead (a line of caulking is called a *bead*) **several times** to make a neat, smooth line. It may seem a bit awkward at first but once you practice on a window or two with a little effort, you'll get the hang of it.

Do not make it too thick or blob it on by squeezing it out too quickly or unsteadily. Make it clean and cover the entire opening. If you use too much it will crack faster and fall off sooner. Also, don't be afraid to wipe it all off and start again. It doesn't take that much additional work. A good caulking job should last between 5-10 years. It should be checked periodically within this time frame. Some tubes guarantee a 25-year life and this is simply not true!

This is a great project. It is easy to do and very inexpensive. If it takes a little longer than expected it is still well worth the effort. The time you spent will be returned in energy savings and lower utility bills.

> This is the same method and the same material you would use in caulking around the bathtub, shower, sink or counter tops. Remember, use the same type of caulk—Acrylic Latex Caulking with Silicone.

Chapter 7

INSULATED GLASS

This is not a step-by-step project nor do I recommend that you do it yourself. If you're remodeling or building a home, this information will be invaluable when deciding to purchase the most energy efficient type of window. If you have an existing home and you are thinking of replacing your present windows, this is for you also. But to try and do it yourself, my advice is to forget it!

"Light and Bright" continues to be the trend in home designs. The days of tiny rooms and tiny windows are now history. Gone and gladly forgotten are the days of the 14' wall with a single 5'x3' sliding window. Today, the common practice is to have two sets of double 3'x6' windows often with arch tops. Instead of 15 square foot windows, we often see 72 square feet or more. High percentages of glass in exterior walls seems to be here to stay.

Window manufacturers are called on to build larger and larger windows to satisfy this desire for "light and bright." There is no question we Americans want our light, our view, and unique window shapes and designs. The problem, however, is that we also want low utility bills, low fading, comfort and most of all, low prices!

To satisfy cost, many manufacturers have gone to

a thinner glass and aluminum frames. This generally is not a problem except in extremely high or low temperatures, in direct sunlight or where strong wind bombards the windows.

Lower utility bills, reduced fading and ultimate comfort are difficult to accomplish. Glass manufacturers continually research to develop ways to control heat, direct sunlight, condensation and conduction properties of glass.

Although great strides have been made in the last ten years, many new developments are on the horizon. New composition of glass, glass that changes colors both mechanically and electronically and new forms of insulation between panes are in the offing. Until they become more in demand, many of these new products will be too expensive for most homeowners. The good news is there are now numerous options a homeowner can choose that are both economical and effective.

Again, the "R" value, which is the "resistance" of heat transfer through the glass, will help determine your windows. There are many items which affect the calculating of the "R" value. The higher the "R" value number the better the insulating qualities. Combine these numbers as an example of the "R" value of the window you are considering.

Type of Window	"R" Value
Single Strength Glass	½
Double Strength Glass	1
Triple Strength Glass	1½
¼" Laminated Glass	1½
Dead Air Space (at least ¼")	1
Low "E" Coating (factory applied)	1
Argon Gas Filled	1

While not normally rated as "R" value, the following are worth including in our rule of thumb:

Type of Window	"R" Value
Small Lower Insect Screen	¼
Full Insect Screen	½
Full Solar Screen	1
Louvered Exterior (open) Shutters	1

Interior items such as mini-blinds, drapes, and window film are not included because, while they are beneficial, they do their work after the window has already been affected by the heat or cold.

An example of a high efficiency glass package would be: Double strength, Low E glass, Argon filled and fixed with a Full Solar Screen. And would add up as follows:

Window Specifications	"R" Value
Double Strength Inside Glass	1
Dead Airspace	1
Argon Filled	1
Low "E" Coating	1
Double Strength Outside Glass	1
Full Solar Screen	1
TOTAL "R" for Rule of Thumb	6

This window is approximately 12 times as effective at stopping the heat transfer as a common single pane, single strength, glass window.

While retrofitting a window opening with the above upgrades can be fairly expensive, buying these during the new construction stage saves money. Many window manufacturers are now offering these features at very low cost when they build the initial window. That's because custom fitting a window in an existing home means the window has to be specially built for that particular opening. Assembly line windows of standard size are far less expensive to manufacture.

Here are a few other features to consider which are also important in choosing glass for your new windows.

✔What does the manufacturer use to separate the panes of glass?

Metal spacers are the least expensive but increase the chance of seal rupture (due to expansion). The best spacers available now are non-conducting separators such as the *swiggle* seal.

✔What is the insulated glass warranty in case of seal ruptures?

Many manufacturers now give a *Lifetime Replacement Warranty.* Glass can be very expensive and is not covered by homeowner's insurance.

✔What about tinting?

Never tint the inside pane of glass, it will void glass seal warranties and often increase dead air space temperatures to dangerous levels.

✔What about Colonial grills?

Colonial grills may *look* attractive but they often decrease a windows' efficiency by as much as half! Colonial grills are the window panes divided into little squares.

✔What do they mean by the "U" value of a window?

Windows are often given "U" values. To convert a "U" value to a common "R" value, simply divide the "U" value into the number 1. (Example: A "U" value of 0.5 equals an "R" value of 2).

✔Does the frame contribute to the energy efficiency of the window?

Although a major portion of a window opening is glass, do not forget the importance of selecting a good, energy efficient window frames. I like a solid vinyl frame because they look good, they never need painting and they don't transfer heat.

If "light and bright" is what you want, make sure the glass you choose will allow for low utility bills, less fading and ultimate comfort.

Chapter 8

SOLAR SCREENS

We talked a lot about insulated glass windows which prevent a tremendous amount of heat gain or loss. In the summer as the sun burns through your windows, it heats the house, fades furniture, carpet, draperies and everything else it illuminates.

Solar Screens can be installed over the windows. They are much more effective than tinting because they shade the window. Tinting is put on the inside of glass. The glass still heats before the tinting does its job which is to reduce direct sunlight and heat gain from coming into the house. When the glass gets hot, it heats up the air in the house from conduction.

There are medium and high density solar screens that stop from 70-90% of the radiation from penetrating and still allow an outside view.

You don't need to have solar screens on each window. Windows facing west and northwest are the best choices to prevent the radiation from the summer sun from penetrating your home.

If you live on waterfront property front solar screens are necessary to reduce *reflective heat* from the sun bouncing off the water.

Solar screens must cover the entire window. Unlike tinting, solar screens allow you to open or close your windows and still derive the benefits of shading. Whereas if a tinted window is opened, the shaded effect is lost on the opened portion.

INSTALLING SOLAR SCREENS

There are all types of solar screens on the market. Kits at the home improvement stores are complete with a frame, screws, turn clips or whatever you need and come with simple instructions. Each screen package is different. The key is getting it nice and tight. Take your time and follow the directions enclosed. Use the full-size measurements of your window taken from the outside of your home.

It's difficult to find a do-it-yourself kit for really large windows, round top windows, octagonal or triangular windows. For such a window, save your own energy and order the solar screen custom made and installed.

You'll save a lot of money on your electric bill when you install solar screens. Even if you have them custom made and installed, the cost is reasonable enough to justify the expense in the long term. The big mistake people make is in getting ALL of their windows set up with solar screens instead of the few primary windows that need it. It's a high price to pay to do it just for the looks.

The *old* type solar screens distorted the view tremendously. They were considered very ugly in colors such as green, red and black. In fact they were so unattractive, many neighborhoods had deed restrictions against installing them. The new ones come in gray and black and are very attractive. They don't look any different from the insect screens on most windows.

Double pane windows do **nothing** to control direct sunlight. Direct sunlight requires a shading device, not an additional pane of glass.

Chapter 9

THERMOSTATS

The older thermostats were the best available at the time. The new ones are simply more accurate. The older ones read the temperature with as much as a 5-6° error. In other words, with your temperature set at say 76°, it might get as high as 80° or 82° before it registers and kicks on your air conditioner. This makes your air conditioner constantly play "catch up" thereby using more energy. The new thermostats now read the temperature with perhaps a 1° difference.

The manual thermostat most people have in their home is mistakenly used as an *accelerator.* For instance, when you're feeling warm, you turn it down low thinking it will cool you faster. The fact is, your thermostat *won't* make the air come on cooler or faster. The thermostat really acts as a brake and when the temperature gets to where you set it. It turns the air conditioner off.

To be wise, set your thermostat at the temperature you feel most comfortable with and wait for the unit to do its job. Turning it all the way down or up won't accomplish anything—your unit works at the same speed all the time.

All manual thermostats have two switches: one switch has an OFF setting for the AIR CONDITIONING or HEATER position and the other switch controls the

FAN MOTOR. This can be set ON or AUTO. Many people make the mistake of leaving the fan on the ON setting.

Your air conditioner has an EVAPORATOR COIL that removes humidity. It's a DE-humidifier that removes the moisture from the air and deposits it in the drip pan below the evaporator coil. By keeping the fan ON while the AC is OFF, the fan picks up this water before it has time to drain. By blowing hot air across the evaporator coil it picks up the moisture from your drip pan and dumps it into your duct work. This causes mold and mildew to grow in the ducts and in your house which is very unhealthy. This is a major contributor of the "sick building syndrome."

Legionnaire's Disease, which kills people, is caused by mold and mildew growing in an air conditioner drip pan.

If you leave the fan ON in the winter, it blows this cold air though the ducts and causes drafts. The ON switch is just there for your service technician to do annual maintenance checkups. In running your system **always** leave the switch on the AUTO position.

There is a new PROGRAMMABLE thermostat that simply changes the temperature at different times of the day. If you live alone and work outside your home, it's wise to have a *programmable thermostat.* You can set it to come on a half hour before you return from work and go off as soon as you leave for work. If you're interested in purchasing a programmable thermostat, look carefully at those available and make certain it has the program capabilities you want.

INSTALLING A THERMOSTAT

This is a simple project. Just follow these step-by-step directions.

1. Remove the existing thermostat by pulling the cover off and removing the screws that hold it to the wall.

2. On the back of the thermostat plate are four wires: red, yellow, green and white. Loosen the screws that hold the wires to the back of the plate and disconnect them. This leaves the four colored wires sticking out of the wall. Trash the thermostat.

3. On the back of the new thermostat, put on your glasses and look for the letters marked on it: R, Y, G and W. Match the wires (R for red and so on) and connect them to the screws.

4. Now attach the new thermostat plate to the wall. If you're mounting it to sheetrock get the new plastic anchors to hold the screws firmly in place.

5. Make certain it's **level**! No need to measure. Just give it a good eye-balling.

6. Snap the new cover plate on and you're ready to roll.

Chapter 10

ENERGY SAVING LIGHT BULBS

We all take electricity for granted. We turn lights *on* at night and *on and off* during the day. We flick the switch *dozens* of times daily and forget what we're doing, how much we're spending and what we can do about it.

I know, you have other things that are more important than to be concerned over the few pennies it must cost to turn a light bulb *on* or *off.* It seems like an insignificant amount of money we're talking about. And, you're right! But when we add ALL of these energy-savings tips together there are substantial savings to be found.

Most people don't know what I'm about to share with you so when the party conversation gets into a lull, amaze them with your newfound knowledge about light bulbs.

INCANDESCENT BULBS

In reality, today's light bulbs are very inefficient. Soon these incandescent light bulbs will be a thing of the past. An entire new range of light bulbs are now available to homeowners.

As of October 31, 1995, the United States Government created a new set of energy standards that light bulbs must meet for certain uses in homes and offices. This is mandated with the Energy Policy Act of 1992 which states, "*Efficiency standards will be established for lamps (light bulbs), motors and distribution transformers. A testing and labeling program is required for luminaries.*" I'll state this in simple language.

Basically, light bulbs will have to meet a set of energy efficient standards and the labeling will have to be explained to the consumer on the packaging. Instead of buying light bulbs based on *wattage*, light bulbs will be based on *lumens per watt*! What this means is you might end up buying a light bulb that uses 15 watts of electricity but gives out as much light as a 60-watt incandescent bulb—and costs five times as much!

I can predict from the packaging of these new bulbs now on the market that the labeling is confusing to consumers. Unless you know what to look for, you'll waste a lot of money.

I don't mean to disturb you, only to make you aware of what's coming so you can, in fact, make the right decision and save a lot of money on your energy bill.

COMPACT FLUORESCENT BULBS

These are the "new" bulbs we will soon all be using. They provide the same amount of light and use a fraction of electricity. Compact Fluorescent light bulbs will soon become the main source of light for the world! They are sorta like people; they come in different shapes and sizes and each does a particular job.

The PRICE of these bulbs will be substantially higher and the initial cost might cause you dismay. But remember, they will end up SAVING you money on your energy bill and will last about 10 times longer than the bulbs we are all now using. At least, that's what the packages say.

These new bulbs are made to fit into your present receptacles. You can put them in lamps, ceiling lights, ceiling fans and recessed lights. But, there are some things you need to know before you buy them such as: **where** they are used and **how** they are used.

Compact Florescent light bulbs need to be put in light fixtures kept on for long periods of time. It's best they stay ON at least four hours at a time. If you put them in a light that you turn on and off regularly, the lifespan of that bulb will be drastically reduced. The advantage of this bulb is they are *reputed* to last up to **10 times longer** than your present bulb—*if* it is used properly!

Do *not* put your compact fluorescent bulbs in a three-way lamp. In some lamps, on the low setting, they tend to sizzle the ballast because they are not getting enough juice. This can become a fire hazard! They are meant for a regular switch—*not* dimmers—and a single type lamp. It they are used in an **outside** light and the temperature drops to zero or below they will not turn on.

HALOGEN BULBS

These bulbs have been around a while and they produce a white, pure light. Most of you have them for your automobile headlights. You know how powerful they are, especially for those motorists who are coming "at" you with their left headlight maladjusted. It causes you to slam on the brakes, shriek rather loudly, and cross your arm over your eyes.

For those of you who live in a temperate to hot climate, a rather mild disadvantage with these halogen bulbs is although they give off this bright, pure light, they also give off much more heat! So, when putting these halogen bulbs in your home, put them in areas where you need high-quality light such as a reading area or, in areas where the lights are repeatedly turned on and off such as hallways and closets.

The old exterior floodlight bulbs have proven to be extremely inefficient are now replaced with the new halogen exterior floodlight. They last longer and give a much brighter light. For instance, a halogen bulb that uses but 45 watts of electricity will give off as much light as a 75-watt floodlight. I suggest you use these on porches, driveways, garden or security lights.

Halogen bulbs can be used in dimmer switches *but* it's important to understand if they are left on low settings for long periods of time their life will be shortened. If used properly, they burn less electricity and have a longer life.

As a step-by-step project with these light bulbs, here is what I recommend. Before you go shopping for new bulbs, walk around your house and look at the lights you tend to leave on a lot. Make a list of the ones to replace with the new compact flourescents. In areas turned on or off a lot or where a dimmer or three-way switch (or outside lights) is used, replace with the new halogen bulbs.

Also on that list, write down the wattage of the incandescent bulbs you have and replace them—not by wattage—but by lumens, with the new, energy efficient light bulbs. If you don't know how many *lumens* are now produced by your incandescent light bulbs, read the package information at the store. As you continue to shop for light bulbs, get accustomed to looking at the *lumens* and not the wattage.

Chapter 11

FIREPLACES

Many people fail to realize the FIREPLACE can be either *energy efficient* or an *energy hog.* In most cases, they take up a lot of energy. It isn't true a big fire in your fireplace will heat up the house. It doesn't really provide much heat at all. It just looks great and sets the mood for relaxation and/or romance.

Fireplaces can be designed to heat a single room, or an entire house. But most fireplaces are simply a fire box with a *flue*, a *chimney*, a *screen,* maybe *glass doors* which simply let the heat out!

This is the way it works in most homes. You have a fireplace and the weather turns cool. You buy and carry in logs to build a fire. You might turn your thermostat down a few degrees anticipating the rise in heat but, the heat goes up the chimney! Heat rises.

You will, however, feel the *radiant heat* from the fire and it is comfortable and warm. If you wander away from the fire cold reality sets in and you rush to turn the thermostat back up a few notches.

With these new fireplaces, those designed to absorb the heat from the fireplace and run that heat through the air ducts from the flue in the chimney—these *can* actually heat your home. However, most homes don't have them.

The cost of a fireplace designed to heat a room or a house, can run anywhere from six to 15 THOUSAND dollars. Whereas the cost of what we might call a "regular" fireplace might be 2 thousand dollars. I personally, love a fireplace to create an ambiance that is relaxing but not as a heating tool.

Here are some things of the inconveniences of a fireplace. First, you are in the middle of your house hauling in wood and (undoubtedly) dropping "stuff" on the carpet. Somebody has to vacuum it. The wood must be purchased and stored somewhere in your yard or garage. This takes up space. Bringing the wood in is always a chore because it's cold and not a lot of fun lugging in several arm loads of cut trees.

Next, you are actually going to build a live *fire* in the center of your living room. I've heard horror stories about how people have literally burned their home to the ground by not using this *non-home-heating-box-of-fire* properly.

But if you or your spouse are romantics and you want to enjoy the hearth, let's learn how to do it safely so you can continue to use it *(and your house)* for as long as you live there.

INSPECTIONS

You only need a good quality flashlight to inspect

your fireplace. Look at what might be faulty. Let's start with the firebox. That's the visible part, the place where logs are stacked, where all the firebricks are. Make certain these bricks aren't loose or cracked and the mortar joints are not in need of repair.

Then, look up into the chimney (using that flashlight) and see if the damper is working correctly. Some people have died by having the damper closed allowing smoke to inundate the house. There are usually two chains or one handle above the firebox which you can see in easy reach if you poke your head inside with that flashlight. One chain opens and the other chain closes the damper. With the lever you just *push* or *pull*.

Also as you look into the fireplace, check the damper to see how much creosote has built up on it if any. Creosote is the black "gooky" residue left from burning timbers. If there is more than a ¼" of this stuff on your damper call a chimney sweep to clean it up.

Cracks or holes in the mortar between the bricks in the firebox can also cause problems. FIREPLACE AND STOVE caulking at your local home improvement store can seal the cracks as good as new. Directions are on the tube. It's easy to do. Put it in the cracks, *squish* it down with a putty knife, and everything will be fine.

Lastly, check the chimney from the roof. Inspect what is called the "crown" or rim. This is the top of the chimney. Make certain it isn't cracked or broken where water can penetrate eventually destroying your firebox.

Take a peek down the chimney to see if there are any tiles are missing or broken.

If you need a cap on your chimney, the little roof which protects the crown of the chimney, do replace it right away. It is inexpensive (less than $20) and will keep the birds and rain out of the chimney. Just measure the opening, take a trip to the store and buy one. They are easy to install and will have specific directions on them. The cap in the drawing above in a heavy, expensive one. There are several caps to choose from. You make that choice. They all work!

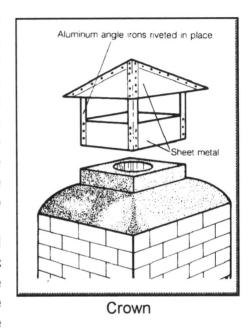

Crown

If you have birds, debris or any pests in your flue, you need to have these obstructions removed. In fact, it is usually less than a hundred bucks to have the chimney cleaned of creosote, checked for loose bricks and mortar, adjustment made to the flue. Your chimney sweep can put a cap on while they're doing the cleaning and remove any foreign objects while he is on the roof.

Some chimneys have wire mesh screens on them

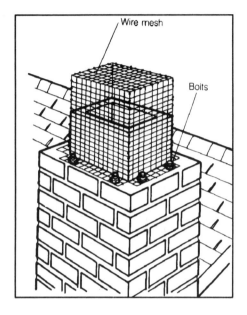

to prevent birds from building nests, or to keep out leaves and other debris blowing in. Another important reason for these mesh coverings is to prevent sparks from coming up the chimney and landing on something combustible and starting a fire.

BUILDING A FIRE

The wood needs to be dry. People say you shouldn't burn pine or mesquite, but you can actually burn *any* wood in a fireplace as long as it's dry. If it's wet, it will smoke and leave inordinate amounts of creosote, the fire-causing residue.

Notice the logs as they are stacked into the grate. They should have small cracks in the ends and feel dry. When storing firewood outside they should be covered with only a roof so the heavy rain will not keep the wood wet.

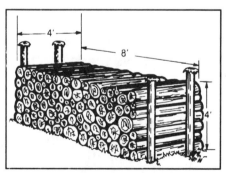

A Cord of Firewood

In STARTING a fire in your fireplace, leave about 1" of ashes on the floor of the firebox itself. This tends to "catch" the coals when they drop which will snuff them out. Also, they pad the big *thunk* you hear as logs slip down or are burned in two.

If you don't have a gas pipe with tiny holes in it to help start your fire, wad up a whole bunch of newspaper and stuff it under the grate. Then add small pieces of dry kindling small twigs or (dry) branches from trees gathered from your yard. **Don't burn plywood or pressure treated lumber** either. The glue in the plywood as well as the chemicals used in pressure treated wood give off toxic gases when burned.

If you have small blocks of 2x4's, or scraps of wood from any projects around the house, save it in a box to use as kindling for starting the fire. Lay the kindling on top of the paper and light the paper. As the paper burns it will ignite the kindling that, in turn, starts your logs burning. I say "logs" but rather *cut* logs; an entire section of a tree trunk will not either start up with this kindling or smoulder rather than burn and is usually too big to fit in the fireplace. Quarter the logs.

Lay your kindling and your logs in a crisscross fashion so air can circulate. To start a fire you need matches, fuel, air and a substance to burn.

When you light the fire, take a piece of newspaper rolled up into a tight bundle like a salami, light the end and push that lighted end up into your fireplace to sort of "condition" your damper. As the hot air rises it *primes* the chimney and prevents getting a down draft resulting in a smoke-filled house. As this paper burns down, quickly toss it under the newspaper you have wadded up under the kindling. The flame should ignite.

Another tip is to crack a window maybe an inch or so open before you start your fire. Many homes that are energy-efficient and built "tight" might tend to have the smoke wanting to come back down the chimney. You've checked the damper is open and "pre-warmed" but the cracked window will boost the oxygen level in the room.

These newspaper logs that you buy at a local store are very clean and burn well. Additional free information is available about fire and fireplaces from:

The Chimney Safety Institute of America
1-800-536-0118

BURNING CHARACTERISTICS OF WOOD

Hardwood	Heat Production	Ignition quality	Splitting quality	Smoke	Sparks	Comments
Apple, ash, beech, birch, dogwood, hard maple, hickory, locust, mesquite, oaks, Pacific madrone, pecan.	High	Difficult	Moderately easy	Light	If disturbed	High quality firewood
Adler, cherry, soft maple, walnut	Moderate	Moderate	Easy	Light	Few	Good quality firewood
Elm, gum, sycamore	Moderate	Moderate	Difficult	Moderate	Few	Good if well seasoned
Aspen, basswood, cottonwood, yellow poplar	Little	Easy	Easy	Moderate	Few	Good as kindling and first logs in fire
Softwood						
Douglas-fir, southern yellow pine	Moderate	Easy	Easy	Heavy	Few	Best of softwoods, but smoke a potential chimney problem
Cypress, redwood	Little	Easy	Easy	Moderate	Few	Usable
Eastern red cedar, western red cedar, white cedar	Little	Easy	Easy	Moderate	Heavy	Usable, best softwood kindling
Eastern white pine, ponderosa pine, sugar pine, western white pine, true firs	Little	Easy	Easy	Moderate	Few	Usable, good kindling
Larch, tamarack	Moderate	Easy	Easy	Heavy	Heavy	Usable
Spruce	Little	Easy	Easy	Moderate	Heavy	Usable, seasoned wood good kindling

Chapter 12

CLEANING AIR CONDITIONING COILS

Contrary to popular belief, the air conditioner does not actually put cold air into your home; it *removes* the heat that builds up during the warm summer months. Of course, the result of this "heat removal" is the cooler air which we very much welcome on hot days.

The component of your air conditioner that removes heat is the **Condensing Coil**. This coil is located on the outside of your home—somewhere. It's the only air conditioning component found outside.

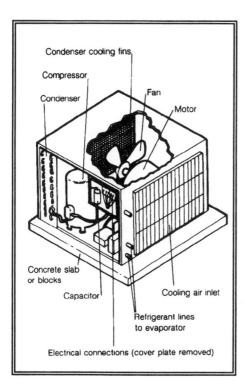

Condenser cooling fins

Compressor

Condenser

Fan

Motor

Concrete slab
or blocks

Capacitor

Cooling air inlet

Refrigerant lines
to evaporator

Electrical connections (cover plate removed)

To give you a better hint of what it looks like, it's just a large metal box or sorts with a motor inside that make all that noise. If you can't see it, **listen** for it. It houses several main components vital to your air conditioner. Each one serves an important function and needs yearly maintenance. The three main components on the condensing unit are:

✔Compressor
✔Fan Motor
✔Condensing Coil

Now, let me explain the functions of each so you might better understand why maintenance will save you money both on repairs and on your electric bill.

The **compressor** resembles a very large can of beans and is usually black. It is located inside the housing near the hose connections that come from inside the house and into the unit.

The **fan motor** is located near the top inside the unit; you'll hear it running.

The **condensing coil** is the largest part of the unit; it surrounds the fan motor and usually makes up three sides of the outside unit.

When the system is operating, notice the fan blowing air across the condensing coil. The air will feel warm to hot. That is the heat being removed from the inside of your house.

If your condensing coil is dirty, it will not remove the heat efficiently and will cost you money in more ways than one. First, it will increase your electric bill and, because it makes the air conditioner run longer and harder, it can cause a breakdown resulting in a service call. That is why you need to clean it every year or get someone to do it. Hopefully, I will explain the procedure thoroughly enough so you can do it yourself.

Step 1. Turn the power OFF to the unit so that cleaning the coil does not become the last job you do. There should be a disconnect next to the unit that is (usually) mounted to the wall of the house and is (usually) gray in color. If you don't see one or are unable to identify it with this description, play it safe and switch the air conditioning breaker OFF in the breaker box.

Step 2. Remove the top of the unit by removing the screws or bolts holding it in place.

Step 3. Using a vacuum or even just your hand, remove all the debris from the bottom of the unit.

Step 4. Being careful not to get the fan motor wet (cover it with a plastic bag if necessary) take the water hose and spray the coil in the opposite direction that the air moves across the coil. Work your way from the top to the bottom washing out the dirt. It's no big chore.

Step 5. Replace the cover and turn on the breaker.

If you feel uncomfortable taking the top off the unit, you can hose the coil off from the outside, again working your way from the top to the bottom. It is not as effective but better than nothing.

This next section is just a money-saving tip I threw in. Most of the time when your A/C unit "acts up" it can be remedied easily and without great cost.

TROUBLESHOOTING YOUR CENTRAL A/C

Trouble	Possible Cause	Solution
A/C does not run	No Power	Check fuses/circuit breakers in service panel
	Thermostat set to "Heat"	
		Reset to "Cool"
A/C cycles on & off or runs continuously	Thermostat set wrong or improperly located	Reset thermostat or relocate it to best reflect house temperature
	Compressor overheating	If compressor cooling motor operates, check if the fan is tight on motor shaft. If motor not operating, replace motor
A/C runs, cools ineffectively	Dirty filter	Replace filter
	Grills clogged, fan or blower dirty	Clean grills, fan or blower
	Belt on blower loose	Tighten belt

Trouble	Possible Cause	Solution
Frost on evaporator coil	Outside temperature too low	Do not operate A/C when outside temperature is below 60 degrees Fahrenheit
	Dirty filter	Clean or replace filter
	Coil fins bent or dirty	Clean anbd vacuum coils; straighten belt fins
Blower motor overheats	Drive belt too tight	Adjust belt tension
	Motor needs lubrication	Oil motor; motors with sealed bearings must be replaced
Water leaks into room or furnace	Condensate drain hole plugged or box and pump cloged	Clean drain, box or pump
A/C excessively noisy	Exterioir grills may be loose	Tighten screws, bolts; use duct tape on panels if necessary
	Blower motor may be loose on mount	Tighten mounting bolts
	Drive pulleys misaligned	Realign pulleys
	Drive belt tension incorrect	Adjust belt tension

Chapter 13

AIR CONDITIONING & HEATING FILTERS

Some of the projects I've addressed in this book are from easy to slightly complicated—but you can do 'em! This next one is the easiest. It is so easy that even my publisher can do it.

A dirty filter can *raise* heating and cooling cost by as much as 10%! If your electric bill is say $200 a month, in a single year by having a clean filter, you can save $220 or more! That is a giant savings from such a little project.

The cost of a filter is minuscule compared to the savings. I've talked with people who live in apartments and rely on their manager to do this. If the manager doesn't do it—*you* do it! Especially if you are paying the bill. Better yet, insist that the manager do it, okay?

DIFFERENT FILTERS

To those of you who listen to my radio program, or watch me on television are familiar with the great disdain I have for some filters which simply don't work.

I'm not apologizing for what I say, nor am I trying to run good people out of business who sell these things. But as a home improvement person who professes some knowledge of what I'm doing, I feel obligated to tell you the truth.

I go to great lengths to test various filters and share the results of my findings—good or bad—with the public. I have personally tried all of them and I have several friends who have tried them all. I talk to a few hundred people each week and receive letters from listeners and viewers. This is my test survey. When an overwhelming majority say a certain item (filter, in this case) is bad or good, I tell you about it and I give you reasons as to why it is either good or bad. It's my job!

Before we get into the *types* of filters, let me explain what a filter does so you will get a clearer picture of why I recommend one type and not the other.

First, a filter DOES not, WILL not, and CANNOT clean the air in your house! People are being sold on this idea and it's an outright, downright LIE! I'll tell you why.

The return air conditioning and heater vent is probably located in the hall outside of one of the bedrooms. There is also a den, living room, *foyer,* kitchen, baths and perhaps a laundry room. The gathering place for most dust is in the laundry room. That's the place where clothes are shaken out, washed, dried and folded. There is simply no way this dust from the laundry room and the rest of the house will travel or can be *pulled* to that area in the hall where your return air duct and filter are found.

For instance, get an old rug and try beating it in the house or your garage. Watch as the dust goes UP and then goes DOWN. NEVER does it start on a horizontal passage toward the air duct or filter. Dust goes up and dust falls down—never sideways. If you had suction strong enough to cause dust to go sideways, you'd need Velcro carpets along with Velcro-soled shoes so you wouldn't be sucked sideways along with the dust. It cannot happen!

The air conditioning and heating filter does one thing: it keeps the dirt in the proximity of the return air duct vent from getting into the AIR CONDITIONING system. When dirt does get in the system (by not changing the filter regularly) the air conditioner will not run as efficiently. It will break down more often and your electric bills will soar!

Now that you know what a filter does, here are the types of filters from which you may choose. Let's begin with the standard filter that most people are aware of:

➔The FIBERGLASS filter that costs a dollar or two. These are the most *inefficient* types but they *are* cheap! We see them stacked everywhere we go, in grocery stores or home improvement centers, hardware stores, everywhere! They are in large boxes that are cut open to show you the colored-filter. They come in many sizes. These need to be changed every 4 weeks.

If you're renting, or simply don't have the money or are too cheap to buy a better filter, use the fiberglass kind;

it's better than nothing. Just be certain to put a new one in every four weeks to help your air conditioner run as efficiently as possible. When it's dirty throw it away and put in a new one.

The filter I like the best is . . .

→The PLEATED filter. It looks like a curtain, it has flexible pleats that can roll back and forth. It has a metal backing, fabric on the front and can be found in either a cardboard or metal frame. They cost from about $4.50 to $7 each and will last up to three months. When they get dirty, toss them and replace them just as you would the fiberglass type.

If you own a condo or a house, please, use the pleated filter. It insures a greater air flow because it's denser and collects 70-90% of all the solids. If you have pets or smokers in your home, you should buy the pleated filter with the CHARCOAL; these are available in many supermarkets and certainly in home improvement stores.

→HI TECH Filter which is an ELECTRONIC filter. These are hooked up to your electrical system and they literally *charge* the particles going through to your air conditioner. These are the best! They still need a pleated filter to keep them working properly and efficiently.

The last filter I plan to mention is the . . .

→The PERMANENT filter, the one I hate. They claim to last forever, cure diseases, allergies and collect and filter the dust throughout the house. It cannot do that! Remember, dust travels up and down, not sideways.

Perhaps someone should invent a NOSE filter or a filter that you can put over your entire head like a beekeepers net. If you change it often enough it will keep the dust and dirt from your nostrils but could prove cumbersome. These filters, regardless of who is selling them is a ripoff. I can't be any nicer than that. It's the truth.

I *really* hate these electrostatic filters. There's a polyester fabric inside them that cannot be cleaned. You see, it's impossible to keep an electrostatic charge in an air filter for any length of time. It wears out and will not recharge itself.

If one room in your home has significantly more dust than the other rooms, I want you to remove the register—that little air conditioning looking grill on the ceiling. You will see how the grill fits into a metal box. If the metal box is not fitted snugly against the sheetrock on the ceiling, or if there's even a little air space in between there, fasten it closed or caulk it (Acrylic latex with silicone). The dust is being pulled from the attic through the tiny cracks.

Even if your room is not gathering dust, take the grills off every year and wash them in the kitchen sink using a water and bleach solution. This should be part of your spring cleaning routine in your home.

Almost every home has grills. They get moldy and

dusty because of the difference in temperature between the air and the grill. Cool and warm air makes the grills sweat which feeds mold and mildew. This is common in most homes and not anything that water, bleach and "elbow grease" won't remedy in minutes. It's not a big problem. Don't let any service people frighten you into thinking something is wrong with your air conditioning and heating system.

Chapter 14

CLEANING REFRIGERATOR COILS

Let's talk about the maintenance of your refrigerator coils. The second largest energy hog in the house other than the air conditioner is the refrigerator. In the context of energy, we can review what the process of refrigeration really does.

The refrigerator removes the heat from inside of this insulated box with shelves and pumps it outside into the kitchen. Then, the heat in the kitchen is removed by the air conditioning system.

To get the heat out of this insulated box, we call a refrigerator, it must go through the coils. In most new refrigerators, the coils are in front, underneath the unit behind the black grill close to the floor. Older refrigerators have coils located on the back.

You've probably had the opportunity to view the more obscure portions of your refrigerator whenever it needed to be moved for installation or repairs. All the dirt and dust clinging in these lower back areas are costing you *big bucks* on operating your appliance. The overload may also cause an early breakdown because of the additional wear and tear.

Coils should be cleaned at least once every six

months. All you need is a vacuum cleaner with a hose attachment and a refrigerator coil cleaning brush. It is a long, thin, round brush you can find for a few bucks at most home improvement centers.

First unplug the refrigerator and remove the black grill (you know the thing you always accidently kick when you get too close to the frig and prop up until later). Use your vacuum cleaner with the special attachment and get down on the floor your side or belly and start brushing the coils. You'll be amazed at what you'll find. I cleaned my refrigerator one time and I thought an entire *dog* was stuck under it and between the coils.

Suck this dirt, dust and debris out (flotsam) with a vacuum cleaner. This will help the life of the refrigerator and cut down the electricity consumption.

After you've cleaned the mess you made, put the grill back on, plug in your refrigerator and turn it back on. Now, check the **gasket** on the refrigerator door. Take a dollar bill (if you're rich, use a 100-dollar bill) and hold it where the door closes, and close the door. If you can slide the bill out between the gasket (the rubber strips in between the door and the frame). If it feels snug, you're okay. If it wants to simply fall out or comes out too easily, the gasket needs to be changed. Try this "bill trick" on the hinge side and all the other sides of the door. You can do this on freezers or refrigerators. If you must have the gasket replaced do just that; *have* it replaced! It is a major chore and more difficult than I want to even think about.

Also, a little two-dollar thermometer can be attached to the inside back wall of your refrigerator to monitor the temperature. It should consistently read between 40°-42° degrees and a freezer between 0°-5° degrees.

Chapter 15

SAFETY TIPS

This is my last chapter on Energy Savings Tips and perhaps the most important. It will save you TIME, ENERGY, PAIN and MONEY on a trip to an emergency room.

Be certain to wear **goggles** when cutting anything. Be careful not to leave electrical tools plugged in when children are around. If an extension cord is needed for outside use (building a fence or tree house) get the "better" heavier kind with a ground outlet.

LADDER

Always use a good quality ladder that is neither damaged nor broken. On a job an experienced roofer friend I once worked with used a nearby faulty ladder in order to save time (maybe 3 minutes) instead of getting his good one from his truck. The faulty ladder broke and the guy can no longer walk. USE GOOD QUALITY EQUIPMENT! It pays high dividends. Here are three safety tips:

★When using a ladder, use it alone. It's not meant to take the added weight of a second person.

★Never go down a ladder backwards (like going down

stairs), FACE IT! Use your legs and arms and ascend and/or descend slowly. Never rush. Those extra few seconds it takes to do this properly could cause you not to fall.

★Always make certain the ladder is on firm ground. Don't place it in mud or on a pile of debris or on uneven ground.

ROOF

Some people simply are not comfortable with climbing or with greater than normal heights. If you're that kind of person, DON'T get on your roof! Even those of you who are experienced and unafraid of heights, can also get into trouble also with complacency. Whether you know it or not there are some problem areas when walking on a roof. Here are two important safety tips:

★Be aware of all the power lines coming to the house that might be close by. Avoid all power lines. They could very well electrocute you.

★It's important to wear soft soled shoes; rubber soled tennis shoes are my first choice. I often see people on a roof wearing work boots or cowboy boots with leather soles. I hold my breath and say a silent prayer for them.

TOOLS

I, being true to my trade, am a *tool person*! I believe in the very best of tools because I use them often. I have learned how to use them correctly. Bad tools can cause a relatively simple job to become a tedious nightmare, use a tremendous amount of time, energy and cause injuries.

I'm not talking about a screwdriver or a hammer but power tools which are so beneficial in getting the job done faster, easier and better. The danger of course is to the person using (or misusing) these tools. Here are some fast tips:

★Always have good, sharp blades on your saw. Dull blades can bind and kick back resulting in injury.

★Always inspect the cords on your power tools to make certain they are not frayed or broken. Exposed wires can electrocute you.

★When working in your shop or on a job site, always plug your extension cords to your power tools into an outlet that is protected by a GFCI (Ground Fault Circuit Interrupter). This will protect you from being electrocuted if there is a bad ground and you happen to be standing in water.

★Golden rule—Never drink and drill. Make certain you're in a good mood and have a cool head. If not, stay clear of power tools. One error in judgment and you can become crippled for life—or die!

And so, my friends, this book must come to a close. I think it's my most practical book because it will save you money. Just about anyone can do most of these jobs to make life easier, less expensive and more comfortable for you (and/or your spouse).

CLOSING STATEMENTS

And so, my friends, this book must come to a close. I think it's my most practical book because it will save you money. Just about anyone can do most of these jobs to make life easier, less expensive, and more comfortable for you and/or your spouse.

My life has been unbelievably fortunate. I'm healthy and my family is healthy. My radio and television career is growing each day and I'm feeling more comfortable with my broadcasts.

I want to thank each of you from the bottom of my heart. It's because of you (who listen to my radio programs, watch me on television and buy my books) that I pursue my chosen field. As my long-winded publisher mentioned in the introduction, I really *do* like to work. The dedication in my first book, *Home Improvement*, may have been a little misleading.

To quote my quotation, "To my children, Jimmy, John and Amanda, If it wasn't for you three I wouldn't work at all," I should have added I love them very much. And, the part that *I work because of them* isn't entirely true; I also work because I enjoy working and helping others.

So, if you continue to support my family by listening to me, watching me and buying my books, I'll do my utmost to keep you abreast of the new items, tools and innovative techniques to make home improvement projects easier for you to save you headaches and money.

I'd like you to know that I will never sell you on a product for personal gain. I'll continue to investigate new items and test them to the best of my ability and will report back the results to you, the consumer.

And a special word for manufacturers. If the product you've *produced* is faulty or misrepresented, I'm going to talk about it to the buying public. To salespersons or store owners. If you're *selling* such a product, don't blame me for putting "thumbs down" on it. If it works, I'll applaud it and if it doesn't, I'm no different than Siskel & Ebert, it's either "thumbs up" or "thumbs down." My job is making sure the consumers get what they pay for and buy only what they need.

Time to get off my soapbox. I just want to say thank you, my fans, once again. Remember, I'm in *your* corner!

Tom Tynan

AN ADDITIONAL FEW TIPS ON SAFETY

* When using a step ladder, open IT ALL THE WAY and do not step on that top rung!

* If you smell gas anywhere around your house, GET OUT and call the gas company from your neighbor's! Even a phone call could spark an explosion.

* In winter, remember to wrap exposed water pipes!

* Other than what I told you earlier in this book about **Ground Fault Interrupters**, ask about them at your local home improvement center. This knowledge could save your life!

* If you're not certain about something while making these repairs, remember it's better to ask *before* and do it correctly. Don't waste time, money or injure yourself!

God bless!

OTHER BOOKS BY SWAN PUBLISHING

HOW NOT TO BE LONELY . . . If you're about to marry, recently divorced or widowed, want to forgive, forget or both, this is an excellent book to read. Candid, positive, entertaining and informative. If you're looking for a new person, this tells **where** to find them, **what** to say and **how to keep them** once you get them. (over 2 million copies sold) . . . . $ 9.95

HOW NOT TO BE LONELY <u>TONIGHT</u> . . . aimed at the *MALE* reader. Other than being courageous and strong, smart women want their man to be sensitive, caring, and understanding. "The" book to give to your man. Or, for men who really want to learn what turns the modern woman on. A fun book for your coffee table . . . . . $ 9.95

NEW FATHER'S BABY GUIDE . . . The "perfect" gift for ALL new fathers. Orientates dad about Lamaze classes, burping, feeding and changing the baby plus 40 side-splitting drawings. Most of all, it tells dad how to **SPOIL** mom! Mothers, sisters, girlfriends, grandmothers, **get this book for dad, he needs it!** . . . . $ 9.95

YOUR FRONT YARD . . . A fun book of information by garden expert John Burrow. It tells about plants, trees, grass, pesticides, fertilizer, lawn and garden maintenance, **everything** you need to know to plant or keep a beautiful front yard . . . . $ 9.95

VEGETABLE GARDENING (Spring and Fall) . . . is yet another fine, fun book written by John Burrow. It tells the size garden you need to feed your family, which vegetables grow best for a country, city, and even an apartment garden on your patio or in planters, *plus,* all about herb gardens to impress your friends . . . . $ 9.95

REVERSING IMPOTENCE FOREVER . . . by Dr. David Mobley and Dr. Steven Wilson, world renowned urologists who specialize in impotence. This book tells it all, and includes diagrams as well as numbers to call for additional information and/or help. It *can* change your sex life for the better. **Women** buy this book 50 to 1 over men. Men, find out what you need to know . . . . $ 9.95

QUEST FOR MEGALODON . . . An adventure book written by ocean-engineer, Tom Dade, about a supposedly extinct shark that ruled the oceans more than 50 million years ago. A physician, a Hall of Fame baseball player and an oceanographer, once boyhood fishing buddies, reunite for the adventure of a lifetime. *JAWS* was 20 feet long and weighed 3 tons. Megalodon is 100 feet long and weighs 60 tons! (About to be made into a movie) $12.95

HOW TO BUY A NEW CAR AND SAVE THOU$ANDS *And how to sell or trade your old one* . . . Cliff Evans, twenty year veteran in sales and management in the retail car business, shares the "innermost secrets" of what car dealerships **don't** want you to know. This information will educate the consumer of sales techniques, financing facts and saving the new car buyer time *and* money . . . .
. . . . $ 9.95

ELVIS IS ON THE LOT, The story of MATTRESS MAC (Jim McIngvale) the owner of Gallery Furniture. He and co-author Dave White tell how a mom and pop retail furniture store has become **America's** #1 single store retailer. Heartwarming, interesting, a great guide for small business. $14.95

* If you can't find these books in retail outlets, write or call us to order by check or credit card. Our address and telephone numbers are on the following page. On books listed above, add $2.90 **per book** for shipping and handling. Expect delivery within 7-10 days.

TOM TYNAN is available for personal appearances, luncheons, banquets, home shows, seminars, etc. He is entertaining and informative. Call (713) 388-2547 for cost and availability.

For a copy of *Vol 1, HOME IMPROVEMENT with Tom Tynan, Vol 2, BUILDING & REMODELING with Tom Tynan, Vol 3, BUYING A HOME & SELLING A HOME by Tom Tynan or Vol 4,* Step-by-Step with Tom Tynan, send a personal check or money order in the amount of $12.85 per copy to:

Swan Publishing
126 Live Oak,
Alvin, TX, 77511
Please allow 7-10 days for delivery.

To order by major credit card 24 hours a day call:
(713) 268-6776 or long distance 1-800-TOM-TYNAN.

Libraries—Bookstores—Quantity Orders:

Swan Publishing
126 Live Oak
Alvin, TX 77511

Call (713) 388-2547
FAX (713) 585-3738